# Rock Picker's Guide
## to Lake Superior's
# NORTH SHORE

*Written by*
MARK SPARKY STENSAAS

*Illustrated by*
RICK KOLLATH

Kollath+Stensaas Publishing
394 Lake Avenue South, Suite 406
Duluth, MN 55802
(218) 727-1731
www.kollathstensaas.com

ROCK PICKER'S GUIDE *to Lake Superior's* NORTH SHORE

©1999 and 2000, 2021 by Mark Stensaas and Rick Kollath. All rights reserved. Except for short excerpts for review purposes, no part of this book may be reproduced or transmitted in any form by any means, electronic or mechanical, including photocopying, without permission in writing from the publisher.

Printed by JS Print Group, Inc. in Duluth, Minnesota
10 9 8 7 6 5   Second Edition

ISBN 0-9673793-0-X

# Acknowledgments

This book would not have been possible without the published research of many geologists who continue to unravel the mysteries of our earth's complex geologic history. In particular we owe a huge debt to University of Minnesota-Duluth professor emeritus Dr. John Green. His expertise on North Shore volcanics and rock identification helped us immensely. Though retired, we know John will continue to educate the public on not only geology but birds and flowers as well.

A big hug to Lisa Sennes for providing support and encouragement on the research trips up the shore. She also gave us a layperson's feedback on the text.

Our reviewers provided valuable feedback and caught mistakes in text we'd already read ten times. So a huge "Thank You!" to Tim Conklin of the University of British Columbia-Vancouver and Dr. John Green.

We'd also like to thank Dr. Susanne Schmidt of the University of Basel, and the original members of the North Shore Climbers Group, especially Dave Pagel, for piquing Rick's interest in geology.

For this second addition, Jane Reed gave us invaluable feedback and assistance. Bob and Nancy Lynch provided us with most of the agate specimens you see on pages 27 to 29. Thanks again J.C.!

Rick Kollath
Mark Sparky Stensaas

Duluth, Minnesota
June 20, 2000

# Contents

Beach Profiles .................................................. 1
   *Corner of the Lake • Kitchi Gammi Park • Stoney Point*
   *Burlington Bay • Flood Bay • Gooseberry Falls State Park*
   *Split Rock River • Beaver River • Tettegouche State Park*
   *Good Harbor Bay • Grand Marais Harbor • Paradise Beach*

Tips for Rock Pickers ........................................ 8
Lake Superior's Geologic Roots ............................ 9
Basalt ............................................................. 13
Ophitic Basalt .................................................. 14
Diabase .......................................................... 15
Gabbro ........................................................... 16
Rhyolite ......................................................... 17
Granite .......................................................... 18
Vesicular Basalt & Rhyolite ................................ 20
Amygdaloidal Basalt & Rhyolite .......................... 22
Porphyry ........................................................ 23
Quartz ........................................................... 24
Lake Superior Agate ......................................... 25
Thomsonite .................................................... 30
Chert and Chalcedony ...................................... 31
Sandstone ...................................................... 33
Special Finds .................................................. 34
Jetsam, Flotsam and Junk ................................. 35
Beaches 101 ................................................... 36
Glossary ........................................................ 38
Index ............................................................ 40
Titles of Interest .............................................. 43

# Beach Profiles

## Corner of the Lake  Duluth on the Lakewalk

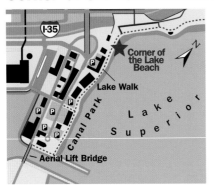

As the Lakewalk passes the hotels of Canal Park and the massive basalt rip-rap (large, piled boulders used as a breakwater), a delightful beach materializes.

Corner of the Lake is a deep rhyolite-dominated beach that shows classic storm ridges. Lakeside the rocks are pebble-sized but farther back on the top of the highest storm ridges you'll find some world class skipping stones. Banded rhyolite can be found alongside solid rhyolite and basalt. Vesicular rocks are absent here but there are some interesting amygdaloidal rocks. The large-grained igneous rocks, granite and gabbro, are very common. Watch also for the "man-made rocks": Concrete, brick, tar and glass "fairy tears."

*Gabbro (see page 16)*

*Rhyolite (see page 17)*

## Kitchi Gammi Park  (Brighton Beach to the locals), Duluth

Look for the Kitchi Gammi Park sign along Highway 61/London Road just northeast of Lester River on the eastern outskirts of Duluth.

The .8 mile-long parkway borders an inviting beach of rocks pea-sized to grapefruit-sized interrupted by outcrops of diabase. Picnic tables, grills and playground equipment make this beach a pleasant lunch stop.

Much of this beach looks red due to the predominance of rhyolite pebbles. Amongst the ubiquitous basalt and rhyolite are some amygdaloidal and vesicular specimens along with their cousins gabbro and granite. Unique here is the vesicular rhyolite with sparkly quartz lining the odd-shaped cavities. Ophitic-textured basalt can be found but is not abundant. This is a decent agate beach with some tan chert and quartz. Also look for banded rhyolite, sandstone, fairy tears, concrete nodules and aluminum blobs.

*Aluminum blob (see page 35)*

*"Fairy tears" are broken glass shards weathered smooth by wave action. (see page 35)*

## Stoney Point  Between Duluth and Two Harbors

*Jasper, a form of quartz (see page 31)*

Stoney Point is 11.3 miles northeast of Duluth's Lester River on Scenic 61. One mile east of the first Stoney Point Drive is a second Stoney Point Drive. Turn in here. The small beach will be a quarter mile in on your left. You can continue on this road to loop past some old fish shacks (relics of the North Shore's commercial fishing heyday) and back to Scenic 61.

Because this stretch of shore faces northeast, has massive rock outcrops and is accessible, it becomes one of Lake Superior's better wave watching sites during storms. The rocks are larger here (no pea-sized gravel) with basalt and rhyolite dominating. Gabbro is more common here than on other beaches. Ophitic-textured basalt is very prevalent. There are some amygdaloidal rocks but no vesicular. Agate, jasper and granite are uncommon.

## Burlington Bay  Two Harbors, on the northeast edge of town

Heading northeast past the last traffic light in Two Harbors, turn towards the lake on the first road after the log Houle Visitor Center. The long sweeping beach of Burlington Bay will be on your left a quarter mile in.

Marble to softball-size rocks populate this shoreline. Basalt, rhyolite, banded basalt, ophitic basalt and amygdaloidal rhyolite all exist peaceably side by side. You won't find any vesicular basalt or rhyolite, though. The microcrystalline quartzes seem to thrive here with a healthy population of Lake Superior agates, yellow chalcedony (unbanded agate), jasper and yellow chert. Fairy tears are also fairly common. Wave-rounded brick fragments masquerade as lightweight yellow rocks.

*Chert from Burlington Bay (see page 31)*

Agate Bay (shown on the map) sounds like a great place to rock pick but is actually the inaccessible and industrial part of the waterfront lined with ore docks.

# 3 BEACH PROFILES

## Flood Bay  Two Harbors. Between milepost 27 and 28

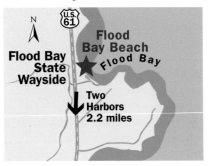

*Rhyolite. The lighter spots are amygdules. (see page 22)*

Just a mile north of Burlington Bay on Highway 61 is Flood Bay. Look for the Flood Bay State Wayside sign. There is an outhouse here but no other facilities.

This is one huge beach and well worth a long search. The igneous rocks are well represented (basalt, rhyolite, diabase, granite, gabbro, amygdaloidal and vesicular basalt). In the quartz family we find jasper, agate, smoky chalcedony, yellow chert and large-grained quartz. Like neighboring Burlington Bay, Flood Bay is home to wave-washed brick fragments and aluminum blobs. Look for the layered flint/siltstone rock called banded flint.

## Gooseberry Falls State Park  Between Two Harbors and Silver Bay at Milepost 39

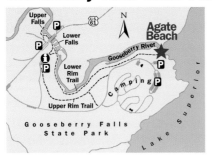

*Granite (see page 18)*

The park entrance is just north of the Minnesota town with the state's absolutely coolest name — Castle Danger. Turn in at the Gooseberry Falls State Park sign immediately after the four-lane ends. You need a Minnesota State Park pass to access the beach. Park at the picnic ground by the lake.

There are several small beaches between the picnic grounds (which is on a wonderful slab of lava flow top) and the mouth of the Gooseberry River. Rhyolite predominates giving the beach a reddish tint when viewed from a distance. Most of the volcanic basalts and rhyolites are solid with no vesicles or amygdules, though an attractive light-banded rhyolite occurs here. This is a decent Lake Superior agate beach so look closely. Granite, ophitic-textured basalt and clear chalcedony are here. Also unique are the chunks of calcite weathered out from the basalt outcrops. Look for a white semi-translucent rock with flat cleavage planes that is easily scratched by a knife. Large chunks can be distinctly rhomboid-shaped (like a 3-D parallelogram).

If you find a rock that could be a stunt-double for a lime jelly bean then you may have discovered a piece of lintonite, otherwise known as the "North Shore emerald" (see page 34).

*Rock picking is enjoyable for everyone so please leave rocks in our state parks for others to appreciate. Thank you!*

## Split Rock River  Between Gooseberry Falls State Park and Silver Bay. Mile post 43 and 44

> Rock picking is enjoyable for everyone so please leave rocks in our state parks for others to appreciate. Thank you!

Four and a half miles north of Gooseberry Falls is the broad sweeping beach of the Split Rock River. (If you come to the Split Rock Lighthouse entrance you've gone too far.) The best parking spots are on the lakeside before or after you cross the river.

There is every size rock here from tiny to too-big-to-carry-home. Ophitic-textured basalt is quite common but vesicular and amygdaloidal basalts are not. Nice specimens of granite, sandstone and quartz have found a home here. If you need a new flint for your flintlock musket, then you've come to the right beach. Large chunks of jasper, which spark readily when struck on a sharp edge with steel, are fairly common. Other microcrystalline quartzes to look for are agate, red carnelian, clear chalcedony and black flint. Wave-washed tar and concrete nodules masquerading as real rocks are probably the result of shoreline erosion and illegal dumping.

*Ophitic-textured basalt (see page 14)*

## Beaver River  Beaver Bay. 1/3 mile past Milepost 51

Just north of the town of Beaver Bay is the Beaver River. Don't miss the gorgeous falls upstream of the road as you cross the bridge. Park on the lakeside just past the bridge. Note the refreshing sign that says *Private Property…Enjoy it…Please leave it clean.* Proceed with caution down the narrow, rutted, steep and sometimes muddy path.

This beach is big, beautiful and worth the trouble. It is a fine beach to picnic on, catch some sun or just take a nap. The muted brown color of the beach when viewed from above is the result of an even mix of the blue-gray basalt and reddish rhyolite rocks. Note the abundant red-striped bluish rock that appears to be basalt but is actually "Silver-Beaver" rhyolite — a locally common variety. Also look for the rare example of porphyry (page 23). Quartz nodules, banded flint and yellow chert represent the quartz family. Rounded rubbery pieces of asphalt will make you do a double take. Now the big secret saved for last… We believe this is the best Lake Superior agate beach on the entire North Shore.

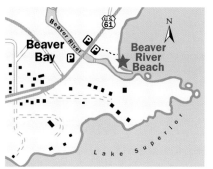

*Silver-Beaver rhyolite (see page 18)*

# BEACH PROFILES

## Tettegouche State Park  Between milepost 58 and 59

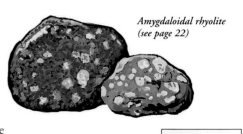

*Amygdaloidal rhyolite (see page 22)*

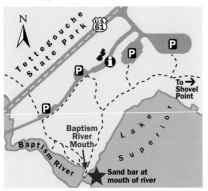

Turn in at Tettegouche State Park and Wayside Rest. Check in at the park office to get your daily or annual Minnesota State Park pass. Continue past the office to the first little parking area just before passing over the Baptism River. Take the trail across the road down to the mouth of the Baptism River.

This is a small beach of larger rocks that the river is constantly rearranging. Side by side with our friends basalt and rhyolite are stones of granite, agate, greenish sparkling diabase and red-striped Silver-Beaver rhyolite (like at Beaver River Beach).

Pretty amygdaloidal rhyolite can be found here. One variety has its irregularly-shaped gas bubbles filled with soft, white calcite while the "peppermint bon-bon" version has its tiny round cavities filled with green and pink minerals.

> Rock picking is enjoyable for everyone so please leave rocks in our state parks for others to appreciate. Thank you!

## Good Harbor Bay  Just south of Grand Marais between Milepost 104 and 105

After passing the Thomsonite Beach Resort, you round the corner of Highway 61 where you get your first view of the coast town of Grand Marais. Turn in at the state wayside. You are four miles southwest of Grand Marais.

Famous to birders for its Long-tailed Ducks (Oldsquaw) in winter, Good Harbor Bay is also noteworthy to rockhounds for its cache of Thomsonite — a semiprecious gemstone. Though Thomsonite Beach is a better place to find the pretty pink and green zeolite, it is in private hands and off limits to all but their guests. Good Harbor Bay is a long arching beach dissected by the low volume runoff of Cut Face Creek. The Thomsonite you are likely to find here occurs as either tiny shards of pink or as resistant nodules in a basalt matrix. But the rocks you'll probably first notice are the hunks of rhyolite dramatically crisscrossed with

*Thomsonite (see page 30)*

*Amygdaloidal basalt and rhyolite (see page 22)*

quartz-crystal-lined cavities, sometimes forming mini-geodes (see left illustration on page 21). Many of the basalts are quite flat and some are ophitic in texture. Agates and porphyry also hide out here. If you noticed the road cut of bedded (layered) reddish sedimentary rock as you rounded the Good Harbor Bay corner on Highway 61, then you saw the parent rock for the beach's abundant flattish sandstone. This sandstone is unusual in that it is made entirely of lava sand grains which were eroded from old flows by river runoff. Later lava flows covered up this deposit and protected the soft sandstone from erosion by the glaciers of the Great Ice Age.

*Sandstone from Good Harbor Bay — an unusual rock to find on the North Shore. (see page 33)*

## Grand Marais Harbor Grand Marais

"Grand Marais," literally translated, is French for Big Marsh but also may have been voyageur slang for a large harbor. The beaches here line both the main harbor and the east side of the peninsula leading out to Artist's Point. The peninsula and the point join to form a tombolo — a geologic term for a beach connecting an island to the mainland. You can easily access the beaches from either the Municipal Campground road (west side) or the Coast Guard/Marina parking lot (east side).

The Grand Marais harbor is home to some of the best skipping rocks on the North Shore. These skippers are almost exclusively basalt and rhyolite with the latter predominating and creating red-tinged beaches. Some of the igneous rocks are porphyritic and some have irregular-shaped amygdules of calcite. Whichever beach you're skipping rocks from watch out for the most abundant beach debris…goose poop.

*Porphyry (see page 23)*

# 7 | BEACH PROFILES

## Paradise Beach  Between Grand Marais and Hovland. Milepost 123

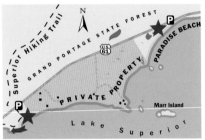

Paradise Beach is 13 miles northeast of Grand Marais and southwest of Hovland. Though the "Paradise" sign at the motel across the highway is gone, this humongous beach is still easy to find. Simply park at the beachside pull-off at the milepost 123 sign and get out.

At least a half mile long with a northeast exposure, Paradise Beach displays classic storm ridges — the lines of rock pushed into parallel ridges by storm waves (see Beaches 101, page 36). Along with basalt and rhyolite are nice examples of granite ranging in color from red to yellow to white. Of the microcrystalline quartzes one can find yellow chalcedony, red carnelian, jasper, banded black flint and a rare form of agate. The Paradise Beach agate is usually tiny and has striking opaque white and orange-red banding. They can sometimes be seen in matrix (still embedded in the base basalt). White and yellowish large-grained quartzes are easy to spot in the wave splash zone.

*Banded black flint*
*(see page 31)*

*Quartz*
*(see page 24)*

# Tips for Rock Pickers

1. Pick after a good storm. The waves toss up a whole new batch of rocks…and possibly a prize agate.

2. Don't ignore late season picking. September, October and even November can have pleasant days in which the beaches are nearly deserted.

3. Check the water's edge where the rocks are wet. Agate, granite, chert, quartz and jasper are all easier to identify when wet. Wading can also reveal treasures overlooked by "dry-footers."

4. Get down and dirty! Sitting or kneeling in one spot forces you to thoroughly search an area.

5. A small gardener's hand rake helps in turning over layers of rocks.

6. Magnifiers help you see subtleties of rocks such as the delicate banding of small agates. A 10-power loupe worn around the neck is best but any magnifying glass will do.

7. Licking a rock will wet it allowing you to see crystal color and structure more clearly (hey, all the COOOOL geologists do it!). If that technique disgusts you, bring a water spray-bottle or simply dip your specimen in the lake.

8. No sunglasses allowed! Tinted lenses alter nature's true colors making recognition and identification of rock types more difficult.

9. Broad-brimmed hats (and to a lesser degree baseball-type caps) shield your eyes from sun while keeping flies at bay.

10. Running shoes or light hikers are ideal footwear for most beaches. Sandals can be uncomfortable and even dangerous on beaches with bigger outcrops which require some scrambling.

11. Wear a small fanny pack or shoulder bag to hold your treasures and keep hands free. A coat with roomy pockets serves the same purpose.

12. Have a plan in mind for the rocks you bring home (rock aquarium, geology display, sauna rocks, etc.). If you have no purpose for them think about leaving them for others to discover.

13. Take a kid along. They're closer to the ground and have extremely sharp eyes. Plus they don't mind getting wet and dirty.

14. Bring a plastic bag for trash. Hey, the beach is doing you a favor so why not reciprocate?

15. A small notebook and pen will allow you to record observations, indicate favorite spots or write your memoirs.

16. Oh yeah…Bring this book along! That's what we wrote it for.

# LAKE SUPERIOR'S GEOLOGIC ROOTS

A long time ago in a land not so far away, a continent tried to split in two. It was 1.1 billion years ago. That's 1,100 million years ago or 1,100,000,000 years ago or .0011 trillion years ago… anyway… it was before you were born. North America was being torn apart. Rifting is what happens when two plates of land move apart from one another. It is happening as we speak. Europe and North America are grudgingly forced two inches farther apart every year by the Mid-Atlantic Rift. When North America tried to split it was along a line from present-day Kansas northeast through Iowa and central Minnesota to the Lake Superior basin (Figure 1). As Matsch and Ojakangas quip in their essential book *Minnesota's Geology*…

> Duluth, Holyoke, Hinckley, Anoka, and Wells thus missed the chance to be seaside cities on the east coast of "Continent West" [and] Prescott, Austin, West Concord, Hampton, Hayfield,

Figure 1. Map showing North America's attempted split 1.1 billion years ago (after Chase and Gilmer, 1973).

and Wangs, would have been part of the Minnesota Riviera on the west coast of "Continent East."

As rifting continued in the present-day Lake Superior basin, lava, formed by the melting of the upper part of the earth's mantle, oozed out onto the ground. If you've ever seen a National Geographic television special on the volcanoes of Hawaii or the flowing lavas of Iceland that slowly bury entire towns, you have an idea of what our area looked like back then… minus the towns. As each flow cooled, the lava hardened into either bluish/black basalt (higher iron content, see page 13) or the reddish rhyolite (higher silica content, see page 17). Holes in this rock were formed by escaping steam, sulphur dioxide and carbon dioxide gasses that didn't quite make it to the surface before the lava hardened (see vesicular basalt and rhyolite, page 20). Sometimes the cavities filled in later with other minerals. These are the amygdaloidal basalts and rhyolites (page 22). The entire North Shore of Lake Superior is basically a string of hardened lava flows.

The flows piled up. As a new rifting episode occurred the land would split apart and pour out lavas over the previous flow. Due to the loss of magma beneath the crust in the rift zone and the accumulated weight of cooled lava flows (basalt and rhyolite), the earth's skin began to sag, forming what would later become the Lake Superior basin. The rifting continued for twenty million years (a very brief period in geologic time). We were left with an immense basin of basalt and rhyolite that was 200 miles wide and nearly five miles thick.

John Green, a University of Minnesota-Duluth professor and world expert in North Shore volcanics, has identified hundreds of exposed flows grouped into six distinct units. He uses the imagery of six giant stacks of pancakes lined up along the North Shore and all tilting towards Michigan. The lava flows vary in thickness from only a few feet to the Palisade Head flow that is nearly 200 feet thick. But as mysteriously as it had begun, the rifting quit and Minnesota lost its chance to become a coastal paradise.

The newly formed valley (known today as the Midcontinent Rift) created a new drainage basin which resulted in the formation of several large lakes, one of which was in the present-day Lake Superior basin. At this time Minnesota was smack-dab on the equator. Rivers draining off high land swept sand and mud into the valley which eventually became thick layers of sandstone and shale that filled up the basin. Over the intervening billion years or so, the Lake Superior portion of the Midcontinent Rift Valley was periodically lapped by warm oceans but never inundated. Sharks swam nearby and dinosaurs roamed the land. Northern Minnesota during this period, was a rolling landscape of well-worn hills punctuated by higher resistant knobs and dissected by numerous river valleys but few if any lakes. And then, just when everything was stable and warm, the Great Ice Age stepped in and altered our state to its current condition…flat, cold and spattered with ponds, swamps, bogs, rivers and one mammoth lake called Superior.

It happened very recently. Our ice age lasted from about two million years ago with the advance of the Nebraskan

*Figure 2. Our ice age lasted from about two million years ago with the advance of the Nebraskan Glaciation to the last retreat of the Wisconsin Glaciation 11,000 years ago (after Ojakangas and Matsch, 1982).*

# GEOLOGIC HISTORY *of the* NORTH SHORE

**Typical cross section of North Shore basalt lava flow**

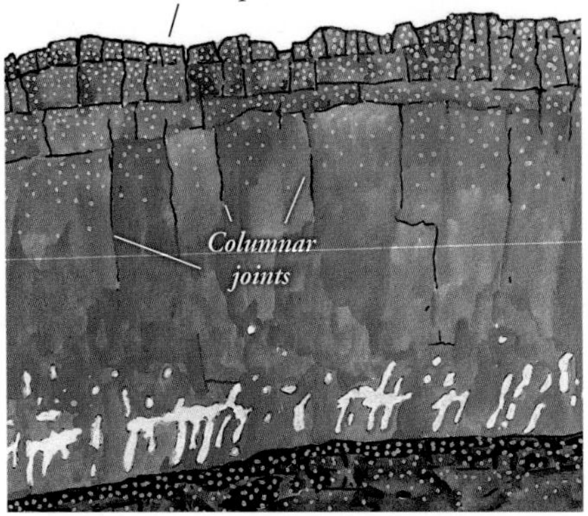

*Figure 3. Bubbles were formed in the lava flows by escaping steam and carbon dioxide gasses that didn't quite make it to the surface before the lava hardened. Beach rocks eroded from the hardened lava with mostly empty bubbles are called* **vesicular**. *Sometimes the cavities filled in with minerals. These rocks are called* **amygdaloidal**.

Glaciation to the last retreat of the Wisconsin Glaciation 11,000 years ago (Figure 2). A proverbial blink of a geologic eye.

The Great Ice Age all started with the polar ice caps and a cooling climate. The snow that fell in the winter piled up higher and higher since the summers were no longer warm enough to melt the snow pack. As snow compacts it becomes ice. This form of ice is considered a metamorphic rock that is created by altering the sediment (snow) by pressure and time. When the ice built up to a great enough depth it started to move and gravity helped it along.

As the glacial fronts crept south out of Canada, lobes of ice were funneled down the Midcontinent Rift Valley through present-day Minnesota to Kansas and Missouri. The soft sandstones and shales were bulldozed clean away by glaciers that may have been 3,000 feet thick or more. The final glacial period between 30,000 and 11,000 years ago saw one glacier, the Superior Lobe, perform the final (for now) gouging of the Lake Superior basin. The crumbling lava flow tops of basalt and rhyolite, as well as amygdules found within them (like agates), were ground up and carried south by the advancing ice. Today, Lake Superior agates are even found in parts of northern Iowa. The rocks you now see on the beaches of Lake Superior not only came from the bedrock of our area but even from as far away as Hudson Bay. After being broken off by these and the preceding glaciers, small as well as large rock chunks were carried far south of their place of origin and deposited on the ground as the ice disintegrated.

Meltwater from the disappearing glacial front started filling the depression and covering up the rocks and till it had left behind. Imagine the scene. Large braided rivers flowed out of the glacial front and over newly abraded bedrock. Huge chunks of stranded ice will later melt to form many of Minnesota's 15,000 lakes. Mammoths, Mastodons, Giant Camels, Saber-toothed Cats and Giant Short-faced Bears were moving into the area as vegetation took hold. At the same time, the retreating glacier was damming the basin to the north as it continued to fill with meltwater and river runoff forming the original Lake Superior (known to geologists as Glacial Lake Duluth). The beachline of the Lake was much higher then. In fact, Duluth's Skyline Drive is a 12,000 year old beach that is today some 500 feet above the current lake level. (Newer and nearly intact old beaches of wave-washed rock can still be found twenty to fifty feet above the Lake.) Lake Superior and the rest of the Great Lakes settled at their current levels about 7500 years ago.

With the weight of the glaciers removed from northern Minnesota and Wisconsin, the land began to rebound. It is still rebounding. But the land is rising even faster in Canada so the water in our end of Lake Superior is getting deeper. Amazingly, in just several thousand years Duluth's Canal Park may become Canal Park Reef! And who knows, before that happens another glacial episode may overtake us. Geologists, after all, call our current period the "Present Interglacial."

So there you have it...a brief look at how we ended up with Lake Superior and the rocks that form her beaches. North America's attempted split gave rise to the magma that cooled to form our gabbro, diabase, rhyolite and basalt bedrock. Glaciers ground down the bedrock and broke off chunks carrying the pieces to our area. Melting ice and rivers filled the basin covering the rocks with water. Lake Superior put the final polishing on the rocks as she rolled them around with other rocks. Her waves threw them up on the beach for you to enjoy.

And finally, we give you this book — a field guide really — to look up the rocks you find. So get out there and enjoy the beaches, cliffs, ridges and rocks of our beautiful and fascinating North Shore of Lake Superior.

*Table 1. Classification of igneous rock types. Nearly all North Shore rock outcrops are igneous.*

|  | Small Crystals (cooled quickly on surface) | Large Crystals (cooled slowly deep underground) |
|---|---|---|
| Iron-rich Silica-poor | Basalt | Gabbro |
| Silica-rich Iron-poor | Rhyolite | Granite |

# BASALT

*Iron, magnesium and calcium give basalt its bluish color.*

*Dry basalt may have a lighter blue-gray appearance.*

*Note the fine grain of the crystals. Gabbro, which can also be blue/black, has larger visible crystals.*

*Wet basalt will have a much darker appearance.*

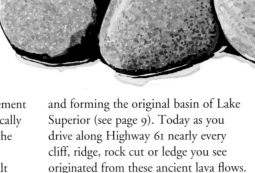

Basalt and rhyolite are by far the most common rocks you will find on Lake Superior's North Shore beaches. Basalt is one of the four rock-types that make up the basement bedrock of the earth's crust. Basically it is lava that cooled quickly on the surface. The faster lava cools the smaller the crystals are. Our basalt originated during the Precambrian era 1.1 billion years ago when North America tried to split right down the middle. Millions of tons of lava flowed out from the rifts depressing the land and forming the original basin of Lake Superior (see page 9). Today as you drive along Highway 61 nearly every cliff, ridge, rock cut or ledge you see originated from these ancient lava flows.

When you look at a smooth bluish-black piece of basalt it is hard to pick out the individual crystals. The lava type that forms basalt is the most common in the world. This is the type you see flowing over roads and engulfing houses in videotape from Hawaii and Iceland. It is rich in iron, magnesium and calcium but poor in silica (50 percent or less) and potassium.

The mile-high glaciers of ten thousand years ago ground down the lava flows breaking them apart and grinding up billions of pieces of basalt. The lake has finished the polishing job and the waves presented them to us.

# Ophitic Basalt

I used to call this "fuzzy basalt" because it appears to be covered with lighter colored fuzzy spots. In actuality, each fuzzy spot outlines a single pyroxene blob that is crisscrossed by minute plagioclase feldspar crystals but no olivine. This crystallization pattern is found in many basalts. Ophitic texturing results from the differential weathering of these minerals. It is NOT the result of filled vesicles forming amygdules.

**How to recognize:**
Bluish-black with very small crystals. Can be found in all sizes and shapes from round to flat. An excellent skipping stone. May be amygdaloidal (gas bubbles filled with minerals, see page 22), vesicular (empty gas bubbles, see page 20), or solid (no gas bubbles).

**Where to find:**
Any and every beach on the North Shore. Careful! You just tripped on one.

**How to recognize:**
Look for the slate-gray/blue basalts that have a mottled appearance. Spots are lighter in color than the rest of the mix and vary in size from pin head-sized to pea-sized.

**Where to find:**
Stoney Point and Split Rock River seem to harbor good numbers of this weird-looking cousin to the straight-laced, ordinary basalt.

*Dry ophitic basalt*
*Wet ophitic basalt*

*The fuzzy appearance results from the weathering of minerals that crystallized in distinct blobs.*

# DIABASE & GABBRO

# DIABASE

*Larger crystals are the result of this igneous rock slowly cooling underground — but not as slowly as gabbro.*

*Unlike basalt, diabase can look sparkly due to reflection of light off larger crystals.*

Diabase is an intermediate rock between gabbro and basalt. As we learned in the previous texts, fine-grained basalt is the result of lava quick-cooling on the earth's surface. Gabbro on the other hand, cooled very sloooowly, deep underground, forming large crystals. Diabase forms from the same magma type as the other two but cools just below the earth's surface. The result is a dark, medium-grained rock.

**How to recognize:**
Dark blue/black but can look greenish and sparkly. Look for a basalt look-alike that has visible crystals. Remember, gabbro has crystals that are so large that you can easily see their flat cleavage planes that reflect light.

**Where to find:**
Any North Shore beach will produce a fine example of diabase. Search Flood Bay and Gooseberry Falls for a striking green and sparkly form.

# Gabbro

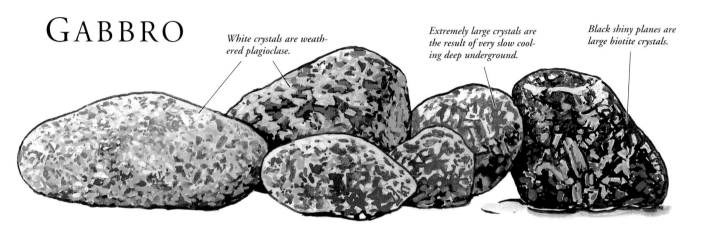

*White crystals are weathered plagioclase.*

*Extremely large crystals are the result of very slow cooling deep underground.*

*Black shiny planes are large biotite crystals.*

Gabbro varies from dark bluish-black to a grayish mixture of very light and darker gray minerals with large and distinct crystals. It is the slowest-cooling cousin of basalt. Magma rich in iron, magnesium and calcium cooled underground allowing time for crystals to grow. It is a beautiful stone for constructing churches, homes and public buildings. Gabbro is not nearly as common as basalt or rhyolite on Lake Superior's beaches.

**How to recognize:**
Black but often noticed only after plagioclase crystals have weathered to white creating an eye-catching black-and-white flecked rock. Occasional specimens may be mostly white due to much weathered plagioclase. Gabbro sometimes shows large crystals of black biotite which have flat cleavage planes that reflect light.

**Where to find:**
The two beaches closest to Duluth, Corner of the Lake and Kitchi Gammi Park, are the best places to pick up a specimen of gabbro. Weathered white-flecked examples abound at both spots.

# RHYOLITE & GRANITE

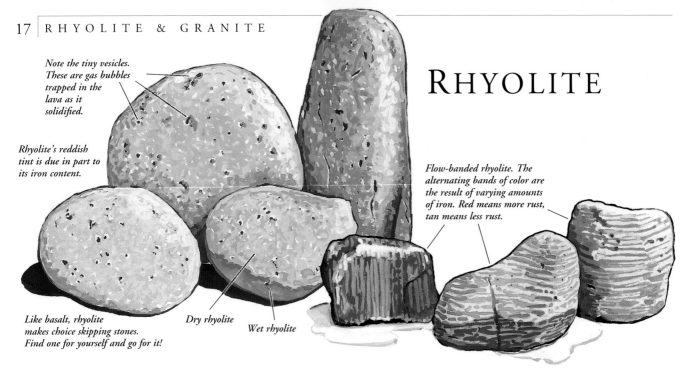

# RHYOLITE

Note the tiny vesicles. These are gas bubbles trapped in the lava as it solidified.

Rhyolite's reddish tint is due in part to its iron content.

Flow-banded rhyolite. The alternating bands of color are the result of varying amounts of iron. Red means more rust, tan means less rust.

Like basalt, rhyolite makes choice skipping stones. Find one for yourself and go for it!

Dry rhyolite

Wet rhyolite

Rhyolite is simply made up of another type of lava that cooled quickly but is composed of different minerals than basalt. It is poor in iron, magnesium and calcium and rich in potassium and silica (70-plus percent). Rhyolite is commonly reddish. Though poor in iron, there is enough to stain the rock a pale red or salmon. Basically, rhyolite is granitic magma that reached the surface and cooled quickly. If it hadn't, it would have cooled underground slowly and formed granite.

Rhyolite can be more resistant to erosion than basalt. The harder rhyolite cliffs of Palisade Head and Tettegouche State Park's Shovel Point are examples of such a flow. Gouging glaciers were unable to level these solid flows.

**How to recognize:**
Usually reddish but sometimes very pale salmon to buff. Hard, smooth and can be spherical or very flat. A good skipping stone. Comes in this sizing spread: XXS, XS, SM, MED, LG, XL, XXL OR XXXL! May be amygdaloidal, vesicular, porphyritic, solid or banded, or most of the above. Silver-Beaver rhyolite is a red-striped bluish rock.

# GRANITE

**Where to find:**

Every beach along the North Shore. But its abundance on the beaches of Kitchi Gammi Park, Gooseberry Falls and Grand Marais tints the whole beach red. Banded rhyolite is common at Corner of the Lake. Silver-Beaver rhyolite is found on the beaches of the Beaver River and in Tettegouche State Park.

*Silver-Beaver rhyolite*

*Wow! This unique rock came from the lava flow that created Palisade Head. Like many rhyolites, Palisade rhyolite is porphyritic, meaning visible crystals formed slowly while the rock was still molten. (see Porphyry on page 23)*

*phenocrysts of quartz and feldspar*

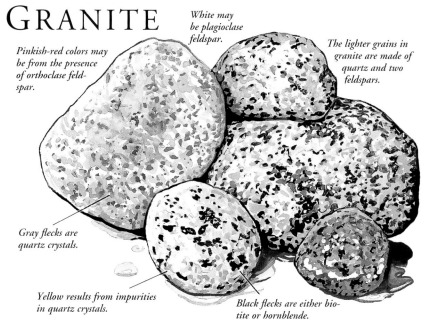

*Pinkish-red colors may be from the presence of orthoclase feldspar.*

*White may be plagioclase feldspar.*

*The lighter grains in granite are made of quartz and two feldspars.*

*Gray flecks are quartz crystals.*

*Yellow results from impurities in quartz crystals.*

*Black flecks are either biotite or hornblende.*

We are all familiar with granite. We see it as polished gravestones, building facades and decorative rock in landscaping. Of course granite is the most common exposed bedrock on the dry face of the planet. (Under the oceans basalt is the most common surface bedrock.) It is what much of our mountains are composed of. The Rockies, Appalachian Mountains, Black Hills and Mount Rushmore, Half Dome, El Capitan and most of the Sierra Nevada, Mt. McKinley (Denali) and the whole Alaska and Brooks Range all have granite as a core component. Closer to home is the Canadian Shield — a huge mass of bedrock (mostly granite) that

# GRANITE & VESICULAR ROCKS

anchors much of Canada, northern Minnesota, northern Wisconsin, Upper Peninsula of Michigan, northern New England and underlies the lava flows that resulted from the Midcontinent Rift.

Granite rocks are not nearly as common on North Shore beaches as basalt or rhyolite. Extremely variable, granite can be predominantly white with black flecks, reddish with black flecks, or other combinations of yellow, black, red, white, gray and pink. The flecks of color are actually large crystals of different minerals. Black is biotite (shiny and flaky) or hornblende (glassy green-black). Gray and translucent equals quartz. Yellow is quartz too but stained with impurities. Pinks, reds and whites are most likely orthoclase or weathered plagioclase feldspars. Plagioclase can also be greenish or yellowish. As we learned with basalt and rhyolite, the faster the cooling of molten material the smaller the crystals. In the case of granite, which has large crystals, the cooling was very slow. In fact, granite cooled and solidified deep underground allowing the crystals to get quite large. Since no granite outcrops occupy the North Shore, our granite pebbles must have been carried here by the last glaciers from exposed Canadian Shield bedrock to the north.

### How to recognize:

Look for speckled multi-colored rocks that really stand out on a beach primarily occupied by the monotone basalts and rhyolites. Colors may be black and white, red and black, yellow, black and white or any combination thereof. It is a hard rock that can be bowling ball-size to marble-size. Granite's colors really jump out at you when wet.

*Granite bedrock forms much of the "Canadian Shield" which underlies the Boundary Waters Canoe Area. All of the granite found on the beaches of the North Shore are pieces of the Shield that the glaciers busted up and carried south to us from Canada.*

*Crystal colors really stand out on a wet piece of granite.*

## Granite Rocks!

Granite is a strong, dense rock without cleavage planes that makes an excellent building material. Beautifully speckled with a variety of colors, granite makes an attractive landscaping rock. It takes on a high-gloss sheen when polished, making it perfect for countertops, building facades and gravestones.

Some of the country's finest granite is quarried in central Minnesota near St. Cloud and Cold Spring.

# Vesicular Basalt & Rhyolite

**Where to find:**
You can find hunks on any beach (hunks of granite that is) but most common at Stoney Point, Flood Bay, Gooseberry Falls, Split Rock and Paradise Beach.

The lava flows we just discussed were filled with gasses that needed to escape. As the lava flowed out of the earth, dissolved gas (steam, sulphur dioxide and carbon dioxide) formed bubbles which fought their way to the surface through the magma. Here's a visual aid to remember this by: What happens when you take the cap off a 2-liter bottle of pop? All the carbon dioxide bubbles rise to the surface, right? That's the fizz in pop. When the pressure in the bottle is released the carbon dioxide becomes gas and escapes into the air. Well, carbon dioxide and steam are the "fizz" in lava. But lava cools and becomes thick relatively quickly thereby capturing some of those bubbles forever. The gas bubbles that harden in the newly formed basalt are called vesicles and are usually near the top of the flow.

*Pipe vesicles are formed at the base of a lava flow.*

*Vesicles are gas bubbles forever trapped in a hardened lava flow.*

*Lake Superior agates formed in gas bubbles like these.*

# VESICULAR & AMYGDALOIDAL ROCKS

(Not quite fast enough to make their escape!) Occasionally elongated bubbles form and are trapped near the base of a lava flow. These are called pipe vesicles and are quite unusual on the North Shore (see the illustration on page 11).

## How to recognize:
Vesicular rocks are basically just holy rocks... I mean rocks with a lot of empty holes. You could say those holes are fossilized gas bubbles. Rocks with filled-up holes are said to be amygdaloidal. That is the subject of our next plate.

## Where to find:
Found on nearly all beaches. Note, though, that some beaches have a much higher percentage of vesicular basalts and rhyolites than others. Why might this be?

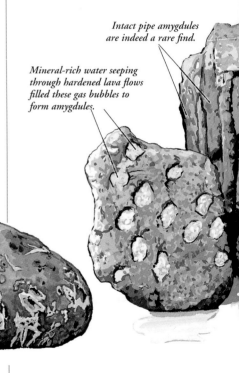

*Intact pipe amygdules are indeed a rare find.*

*Mineral-rich water seeping through hardened lava flows filled these gas bubbles to form amygdules.*

*Fractures in rhyolite lava flows were caused by ancient earthquakes. Cracks filled in later with quartz, calcite or zeolite minerals.*

*Quartz-lined cavities form mini-geodes.*

*Rhyolite*

*This highly fractured rhyolite probably came from flows at the base of Palisade Head.*

# AMYGDALOIDAL BASALT & RHYOLITE

*Calcite, a soft mineral that easily scratches with a knife, is a common filler of gas bubbles.*

*Basalt – note the bluish-gray coloration.*

Read the previous page on vesicular basalt and rhyolite to find out how gas bubbles become trapped in basalt and rhyolite. After the lava cools, forms rock and is buried by the next lava flow, ground water and water associated with the lava percolates down through cracks in the slab. This water, rich in dissolved minerals, fills up the bubbles.

Slowly over time, the water evaporates leaving only the crystallized minerals to occupy its new "fossilized bubble house." Amygdules can be filled with Thomsonite (page 30), agate (page 25), zeolite (soft, pink and white, radiating fan-shaped, page 30), calcite (easily scratched, white), quartz (harder than calcite, usually translucent or gray, can form crystals that line the vesicle, page 24), epidote (hard, grass-green, elongate crystals, usually tiny amygdules), prehnite (hard, very pale-green, lumpy crystalline aggregates, sometimes flecked with copper, page 34) and chlorite (soft, dark green, lines and fills cavities).

### How to recognize:

"Spotted" rocks, you might say. But for it to be truly amygdaloidal the spots must be actual cavities filled with one or more of the above minerals. A dizzying array of varieties can be found. Beware of porphyritic rocks whose large crystals can look like filled gas bubbles (see page 23).

### Where to find:

You can find many varieties of these rocks on most any shoreline. Pipe amygdules are difficult to find on any beach.

# PORPHYRY

A well known landmark and a massive example of porphyritic rock is the United States' first national monument, Devil's Tower in Wyoming. It formed as a giant molten plug which didn't quite make it to the surface of a volcano. The volcano eroded away, leaving just the plug.

Porphyry is unique in that phenocrysts (whole crystals) of feldspars, quartzes or other minerals formed in still-molten magma before it erupted onto the surface. When the rock spewed out of the earth as lava, the matrix (the base rock; the stuff which was still molten while the crystals formed) hardened rapidly, forming a fine-grained rock trapping the large-grained phenocrysts. On the North Shore, the matrix rock can be either basalt or rhyolite.

**Porphyry is not the same as amygdaloidal rock!** In amygdaloidal rock the base rock hardens first, trapping gas bubbles into which minerals then seep and crystallize (vesicles, see page 20). In porphyry, the crystals were already there when the surrounding magma hardened to rock. Some rocks have both. The bulk of Palisade Head and Shovel Point were formed from a single, massive rhyolitic lava flow (the biggest one on the North Shore). If you look closely at rocks from this flow you'll see small, rectangular, pink-tinted crystals of orthoclase feldspar and globby grayish-white crystals of quartz. In the same chunk there may be smaller or larger vesicles, some filled with soft calcite.

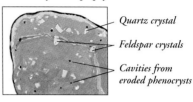

*Palisade rhyolite: a porphyritic rock.*

- Quartz crystal
- Feldspar crystals
- Cavities from eroded phenocrysts

### How to recognize:
You can easily tell phenocrysts and amygdules apart. Porphyritic phenocrysts are boxy and have angular, squarish corners and edges. Amygdules form in bubbles so look for roundish, mineral-filled shapes.

### Where to find:
Rhyolite is everywhere and often porphyritic. Basaltic porphyries are much less common but represented on nearly every North Shore beach.

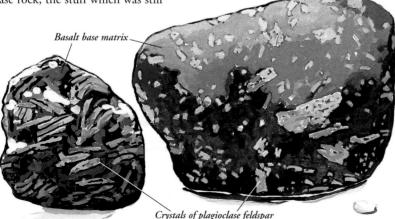

*Basalt base matrix*

*Crystals of plagioclase feldspar*

*Yellowish color is due to impurities in the quartz.*

IMPOSTER ALERT! *A quartzy-looking rock with a dimpled surface is actually an entire amygdule forced from a vesicle whole. It may be agate or chalcedony.*

# QUARTZ

Quartz is one of the most common minerals in the world and a constituent in many rock types (agate, granite, flint, chert, jasper, sandstone and more). It forms under every conceivable condition. It lines cavities and fissures, creates geodes and forms solid veins (look here for gold!). Pure quartz crystals are clear and six-sided though over a thousand crystal forms have been described. Quartzes come in a rainbow of colors from pink (rose quartz) to purple (amethyst).

## How to recognize:

Occasionally found as a whole large-crystalled pebble as pictured here. Beach quartz can be translucent to cloudy and white to yellowish but is almost always quite small. Note that it is not as glossy shiny as chalcedony due to its larger crystal structure.

Quartz is more common as a mineral-filling in cracks and bubbles. But don't confuse it with calcite which is also whitish but much softer and has flat cleavage planes. Quartz cannot be scratched with a knife but calcite can. Calcite breaks up too readily to occur as wave-washed nodules. Watch for purplish pebbles. These may be pieces of amethyst carried south from regions around Thunder Bay, Ontario.

## Where to find:

Quartz pebbles can be found at Kitchi Gammi, Flood Bay, Beaver River and Paradise Beach.

# Lake Superior Agate

A pitted, lustrous surface is actually the outside of the agate nodule. It is an impression of the cavity wall in which it was formed.

Red bands are the result of the rusting of iron within the agate when exposed to oxygen.

The typical concentric ring pattern on agates is called "fortification banding."

The largest Lake Superior agate ever found in Minnesota may be the 108 pound behemoth that is proudly displayed in the lobby of First National Bank in Moose Lake.

As we learned earlier, hundreds of volcanic episodes burped up many megatons of lava. As this lava cooled, gas bubbles (steam and carbon dioxide) formed in the lower levels and tried to escape to the surface. Some bubbles didn't quite make it to the top before the lava became basalt or rhyolite. Further cooling caused contracting which created cracks (say that ten times fast!). Waters rich in minerals were then able to seep into the "petrified bubbles." (Geologists call them vesicles. For more information on vesicles see page 20.) This water came from rain, underground reserves or from the volcanic magma itself. Each band in an agate represents another flush of water with a differing mineral composition, temperature, pressure and time period. Red bands indicate waters richer in iron. On the whole, most agates are over 99 percent microcrystalline quartz — a quartz with extremely tiny crystals. Occasionally, you'll find an agate whose center filled in or partially filled in with slow growing larger crystals of quartz. And sometimes — very rarely — you'll find the fill tube through which the mineral solutions entered the cavity (Figure 4).

We are lucky that agate is harder than basalt or rhyolite. Why? Because if it was softer, the gargantuan grinding glaciers of the Pleistocene would have not only knocked them out of the basalt but ground them to dust in the process.

*Figure 4. Fill tube in a paintstone agate.*

Fortunately for rockhounds everywhere that wasn't the case. What the glaciers did do, though, was to transport some Lake Superior agates well south of the Lake Superior basin. In fact, many were stranded in southern Minnesota and even Iowa when the glaciers retreated. (Who said the only good thing to come out of Iowa is Interstate 35?) Not only did the glaciers transport the agates but they also performed as giant rock tumblers, smoothing the rock's rough edges. Agates on the North Shore were either brought south from Canada or weathered out of our bedrock naturally. Only after the agate is free of the base rock can it be abraded and exposed to the air causing the iron-rich banding to turn red due to oxidization. (In other words…rusting.)

### How to recognize:

Easiest to spot when wet. Agates will glisten in the sun. Beach finds are rarely bigger than a walnut and most are pistachio-sized or smaller. Finding them becomes second nature with practice. The alternating red and clear or white bands are distinctive. They are translucent and extremely hard. Even unbroken agate nodules can be easily identified by their lustrous, dimpled surface.

### Where to find:

Almost any beach with small pebbles will hold agates and probably a prize or two. Remember, the waves of Lake Superior keep replenishing the stock so keep looking — especially after storms or early in the spring. A fist-sized three-pounder was recently found at one of the beaches featured in this book. (Psssst! Here's a secret. The best and biggest Lake Superior agates are found on the dirt roads and gravel pits of Carlton County, Minnesota. This is compliments of those bulldozing glaciers that broke them out of the North Shore lava and roughly escorted them south.)

---

## *Kids Only!*

Agates are like the fudge and walnut chunks in Ben & Jerry's Chunky Monkey® Ice Cream. The chunks are harder than the stuff around them, right? Agates are harder than the basalt rock that surrounds them.

Agates were born as gas bubbles in cooling lava. The gas bubbles are like the bubbly fizz in a glass of pop. When the lava cooled the bubbles became petrified. Then water with lots of stuff in it trickled down and filled in the bubbles. The minerals in the bubbles got hard and made agates. The red stripes are from water with a bit of iron in it that's rusted.

Then the mile-high ice glaciers came and broke the softer lava rock apart and out came the agates for you to find. Some even got pushed down to Iowa!

# Agate Oddballs

**Brecciated Agate**

Chunks of broken agate cemented together by a second flush of silica solutions. This is different from a ruin agate in that this agate was broken to pieces and not merely cracked.

**Crystal Impression Agate**

A rare but fascinating agate phenomenon is the impression of a well formed crystal in the shell of an agate. Apparently the agate formed in a cavity already occupied by single or multiple crystals. The embedded crystals–usually calcite, quartz or feldspar–eroded away long ago.

**Geode Agate**

Occasionally the flow of solutions into a gas bubble stops before a complete agate is formed. The result is a hollow center lined with clear quartz crystals. Rarely a purplish quartz called amethyst fills the void. Geode agates are strong evidence that agates do fill from the outside in with distinct flushes of mineral solutions.

**Eye Agate**

An eye agate has one to dozens of round "eyes" on its surface. (A one-eyed specimen is called a "cyclops.") The eyes are perfectly round and composed of concentric rings of color. How they formed is a big mystery. One idea is that a thick solution flowed into and then out of a gas bubble leaving jelly-like beads on the wall of the cavity. The beads solidified and future flushes of solution crystallized around the existing beads creating the rings of the eyes. A rare find.

*Eye agate*

**Gray or Black Agate**

These subtle yet striking gray-and-white banded agates are regular Lake Superior agates but lack oxidized iron. In other words, they have not yet rusted. Think about it: the most colorful agates are really just full of rust.

**Paintstone Agate**

The deep rich colors of these agates almost seem to be painted on. Saturated reds, blues, greens and yellows are probably caused by high concentrations of minerals in the microcrystalline quartz which oxidize after prolonged exposure to the elements (see Figure 4, page 26).

**Paradise Beach Agate**

Paradise Beach agates are usually small with bands of opaque white and orange. The orangish, painted quality is diagnostic of this "made in Minnesota" rock. They only formed in a few lava flows northeast of Grand Marais. It is a variety of paintstone agate. Look for them at Paradise Beach.

*Paradise Beach agates*

## Peeled Agate

Peeled agates resemble an orange with part of its skin peeled away. That isn't a bad analogy actually. Layers of the agate have broken away along the banding planes. What causes this? Simply the hard knocks of life as a billion-year-old agate. Freeze-thaw cycles, glacial action during the Ice Age and banging around in rivers and on beaches all take their toll.

*Peeled agates*

## Ruin Agate

Offset bands like a mini fault line distinguish this fractured and recemented agate. Eons ago, an earthquake — common in zones of volcanism such as the North Shore a billion years ago — cracked the agate while it was still in the basalt. Later flushes of silica-rich solutions sealed up the crack.

*Detail of a ruin agate. Note the offset banding.*

## Sagenite Agate

Radiating needles of a foreign mineral form a fan-like pattern running through the agate. Sagenitic patterns are best seen on polished specimens. Unbanded or banded and very difficult to identify in the field.

## Shadow Agate

Peering into this translucent agate and then slowly rotating the rock reveals a deep "shadow" inside. Fascinating, but hard to illustrate. Best observed on polished agates.

## Skip-an-atom Agate

Once believed to be an opalized agate, the gray to lavender color is actually due to a diluted combination of bluish-white chalcedony and hematite. The name skip-an-atom is a misnomer based on the assumption it contained opal (which would have had an *extra* water molecule anyway. No obvious banding. Found at Flood Bay.

### Agates for sale! Get your agates here!

Tried your hand at agate picking but came up empty handed? Well all is not lost. If you're in Two Harbors visit Bob and Nancy at Agate City. They are some of the most knowledgeable agate people in Minnesota and their agate selection is spectacular. Up the shore in Beaver Bay you'll find Keith Bartel's Beaver Bay Agate Shop. All sizes of Lake Superior agates are for sale. Check out their world class museum of agates and other Minnesota minerals [beaverbayagateshop.com].

For cyberspace beachcombers try the internet. Three sites that have Lake Superior agates for sale are www.agatecity.com, www.fairburns.com (search under "lakers") and www.amazingagates.com. Remember, there are hundreds of varieties of agates found in the world so search specifically for "Lake Superior agates." Happy virtual rock picking!

# AGATES & THOMSONITE

## Stalactite Agate

These rare agates appear to have been formed when stalactites of silica gel hanging from the roof of the cavity were surrounded by successive flushes of solution forming the banding layers.

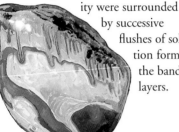

*Stalactite agate (Polished specimen)*

*Waterlevel agate with quartz center and some fortification banding.*

*Waterlevel agates*

*Banded flint*

*Caution: black and gray banded flints can be mistaken for waterlevel agates*

## Tube Agate

On the surface, a tube agate can look like an eye agate. But on broken or polished specimens look for the tube-shaped structures running down into the agate. Tubes formed when chalcedony crystallized around thin rods of minerals suspended in the cavity.

## Waterlevel Agate

Horizontal banding, instead of the usual concentric rings, is the telltale sign that you have a waterlevel agate. Like all agates, they formed in gas bubbles in the lava. But in this cavity, gas was still present so the heavier silica-rich solutions flowed in and settled to the bottom instead of coating the whole cavity. Each layer crystallized before another influx of solution.

## Waterwashed Agate

As the name implies, these well-rounded specimens were likely rolled around in a river or the surf of an ancient beach till smooth. Waterwashed agates are nearly round and less translucent than other agates.

*Waterwashed agate*

### Take your agate to the beauty parlor

Agates look best when wet. Unfortunately, they lose some of their luster as they dry. Displaying agates in a glass bowl with water is fine but the water will soon turn cloudy. Here is a method for making your finds look lustrous year round. Heat the oven to 250 degrees and then turn off. Spread your rocks evenly on a cookie sheet and put in the oven. When they are very warm to the touch take them out. Rub them with pure mineral oil. This oil will seep into the agate's pores, absorbing it evenly. Mineral oil dries thoroughly and does not go rancid. Note how the delicate banding now stands out. Your prize agates are now ready to display. This procedure also works well on jasper, flint, chalcedony and chert.

# THOMSONITE

*Pink zeolite material does not have the concentric rings like Thomsonite but is a close mineral cousin.*

*Unlike agates, Thomsonite is often found still encased by the surrounding basalt bedrock.*

*Surrounding basalt*

*Thomsonite amygdule*

*The radiating bands or concentric circles of pink and pistachio green are unmistakable.*

Formed like agates through water percolation into petrified bubbles in the basalt, Thomsonite is the much rarer of the two. It is even considered a semiprecious stone by some. Jewelry made from the pink, white and green nodules can be spectacular. Geologists Rapp and Wallace called the nodules of zeolite "little bloodshot eyes." An apt name for the rare whole amygdule with an "iris" of pale green and radiating circles or "veins" of pink.

One specific lava flow spawns most of our Thomsonite. It is the 160-foot thick Terrace Point basalt flow and it is the same formation that forms the "teeth" of the Sawtooth Mountains. This is seen most easily from Grand Marais looking southwest down the shore.

**How to recognize:**

The radiating bands or concentric circles of pink and pistachio green are unmistakable. Whole nodules are very rare. More common is a wave-washed piece with hints of pink or green. Still more common is seeing the amygdules still "in matrix" in outcrops of basalt.

**Where to find:**

Good Harbor Bay, a few miles southwest of Grand Marais, is your best bet. The appropriately named Thomsonite Beach, just southwest of Good Harbor Bay, is only accessible to guests of the Thomsonite Beach Inn & Suites. Don't miss their awesome display of fine jewelry and specimens.

# CHERT & CHALCEDONY
## FLINT, JASPER, CARNELIAN

*Jasper is actually a type of quartz with microscopic crystals.*

*Mary Ellen jasper is a variety laced with swirls of petrified Precambrian fossil algae in structures called stromatolites. It was first found in Minnesota's Mary Ellen Mine.*

*Jasper varieties*

*Red speckles are iron-rich hematite sands.*

*Banded jasper*

Most of us are familiar with quartz—the clear large crystalled rock profiled on page 24. Now we move into tiny-grained quartz rocks known as microcrystalline quartzes. These rocks were formed from silica that was dissolved out of silica-rich minerals like quartz. The silica travelled as fine particles in water until settling as a gel in cracks and bubbles (vesicles) in the bedrock. Slowly, the water evaporated leaving only the microscopic quartz crystals of silica.

All microcrystalline quartz can be divided into two groups of rocks: chalcedony (translucent and glossy) and chert (opaque and waxy).

### How to recognize:

Chalcedony rocks are nearly pure silica ($SiO_2$) resulting in a translucent stone with a glossy to waxy surface. Examples include Lake Superior agate (page 25), red carnelian and yellow chalcedony. Cherts, on the other hand, have some impurities causing them to be less translucent and less glossy.

*Wet flint can appear jet black.*

Though duller and opaque, cherts still come in a huge variety of colors from black (flint) to tan, yellow, gray and red (jasper). Jasper is stained with deep-burgundy sands of hematite (iron-bearing rock). With a little experience even tiny jasper nodules become very noticeable to the observant beachcomber. Their deep red/purple sheen is especially visible in the wavesplash zone at the Lake's edge. Jasper may be "freckled" or wavy banded with varying shades of red and yellow. Chipped edges form half-moon shaped divots. This is called conchoidal fracturing and is a characteristic of all microcrystalline quartzes.

**Where to find:**
Let's talk cherts first. Jasper is most plentiful at Split Rock. Of the other cherts, tan is found at Kitchi Gammi while yellow shows up on the beaches of Burlington Bay, Flood Bay and Beaver River. Black chert, a.k.a. flint, is found on the Split Rock River Beach while banded flint is possible at Beaver River and Paradise Beach.

Four colors of chalcedony are out there to be discovered. Look for the clear form at Gooseberry Falls and Split Rock, yellow at Burlington Bay and Paradise Beach and smokey-gray amongst the basalts and rhyolites of Flood Bay. Eagle-eyed rock pickers may spot carnelian (red chalcedony) on Split Rock River Beach or Paradise Beach.

*Carnelian is translucent.*

## Arrowheads & Gunflints

All cherts, including jasper, can be flaked into tools like hide scrapers, spear points and arrowheads. Their microscopic crystal structure allows them to be worked into razor sharp edges. Paleo-Indians to the Ojibwa all used jasper in this fashion. Later, flintlock guns were fitted with pieces of microcrystalline quartz. When the trigger was pulled the flint, jasper, or chert slammed into a steel plate throwing sparks into the pan holding gun powder which fired the musket.

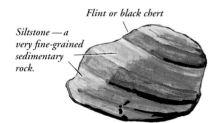

*Siltstone — a very fine-grained sedimentary rock.*

*Flint or black chert*

*Banded flint is sandwiched layers of the glossy black microcrystalline flint and a fine sedimentary rock, siltstone. Since flint is harder than siltstone, it often erodes slower and juts out beyond the siltstone. The rusting of its iron content often leaves a yellowish-red stain.*

*Yellow chert*

*Tan chert*

*Half-moon shaped chips are due to conchoidal fracturing — an identifying trait of microcrystalline quartzes.*

# SANDSTONE

*Look for bedded, horizontal layers when identifying sandstone.*

*Note sandstone's rougher sandpaper-texture compared to rhyolite's smooth surface.*

*Sandstone is a rare rock on North Shore beaches.*

During the period of oozing volcanism that laid down the hundreds of lava flows which formed our rhyolite and basalt, were several breaks in the action. In these times of volcanic inactivity, streams formed from rain that fell on the land. Flowing over the immense lava plain, the water eroded its surface and left sediments which solidified to form beds of sandstone. The sandstone that formed is unusual in that it is mainly composed of bits of volcanic rock. Plagioclase, pyroxene and magnetite minerals are the main ingredients. But much of the sandstone that filled the rift basin after the volcanism stopped consists almost entirely of typical sand which is composed of pure quartz fragments.

### How to recognize:

Sandstone rocks are reddish to tan and very soft. They have a fine-grit sandpaper-type surface. Look for distinct parallel bedded layers. Rhyolite, which is also reddish, is hard, smooth and shows no layering. Pancake-size is large and Oreo®-size is small.

### Where to find:

Nearly exclusive to the beach at Good Harbor Bay. Note the road cut along Highway 61 that exposes a beautiful sandstone outcrop layered between the lava flows.

# Special Finds

## Anorthosite

An intriguing component of the North Shore geology is an odd rock known as anorthosite. Small to gigantic, unmelted blocks of this material floated up from deep in the earth's crust in the North Shore diabase magma until the whole mixture congealed into solid rock. Many of the hilly knobs and domes of the central North Shore are made of this hard and erosion resistant stuff. Split Rock Lighthouse sits on a chunk. Carlton Peak and the hills of Tettegouche State Park are also anorthosite.

Anorthosite is unusual because 98-percent of it is just one mineral: plagioclase feldspar.

*Anorthosite* Shoreside finds will have a creamy, greenish, or gray color and appear translucent and vaguely crystalline. Look closely and you'll see crystal surfaces which reflect light as you rotate your rock. With luck, you'll see tiny parallel striations, or lines, called *albite twinning*.

## Lake Superior Copper

*Copper nugget*

Raw nuggets of copper are rare on the North Shore of Lake Superior but are found regularly on the Keewenaw Peninsula shore of the Upper Peninsula of Michigan. Copper was deposited in the vesicular or broken tops of lava flows.

Minnesota does have native copper deposits. From 1863 to 1865 one ton of the metal was mined out of rock from Knife River. Even so, any beach nugget you might find was probably carried here by the glaciers from Isle Royale.

Native peoples of 5000 years ago treasured the soft malleable metal. It could be cold-hammered into spear points and other tools then traded far and wide. Lake Superior copper tools and jewelry have been found at prehistoric sites from Montana to New York to Georgia.

Look for oxidized-green or copper-colored nuggets.

## Lintonite— North Shore Emerald

*Lintonite*

Jelly beans of gray and green. Geologists do not recognize lintonite as a distinct mineral since it is really just a form of Thomsonite. But to the rock picker it is very distinct. The somewhat translucent pale green pebbles are literally jelly bean-size. Lintonite does not exhibit any banding like Thomsonite, nor does it come in pink. Check Good Harbor Bay.

## Prehnite

Prehnite is formed in basalt gas bubbles just like agates, Thomsonite and other zeolites. Note the mint green to whitish luster and globule-like clusters. It can also be pink due to microscopic specks of copper. More resistant than the base basalt, prehnite amygdules form knobby projections on beach rocks. Isle Royale "Thomsonite" is actually prehnite.

*Prehnite*

# Jetsam, flotsam & junk

Besides the usual garbage that washes up on our beaches, there are some interesting pieces of flotsam and jetsam and junk that make their way in to shore.

**WOOD:** Wave-rounded chunks of wood can look for all the world like a piece of sandstone…until you pick them up and feel their light weight and texture.

**TACONITE PELLETS:** Your first clue that these are not natural rocks is their uniform roundness. Most likely they have fallen off ore boats or loading docks. After iron-ore bearing rock is removed from the ground of Minnesota's Iron Range, it is pulverized to a dust. Huge electromagnets pull magnetic ore dust out from the rest of the rock. It is then mixed with bentonite clay and rolled into marble-sized pellets. After being baked, they are shipped out of Duluth, Two Harbors and Silver Bay to the steel making centers of the eastern U.S.

**GLASS:** Also known as "fairy tears," these wave-washed bits of glass hold great attraction for kids who think they've discovered emeralds or diamonds. White glass is most common. Cobalt blue is the rarest. Green and brown are in between in abundance.

**ALUMINUM BLOBS:** These melted blobs of aluminum are very lightweight. Where do they come from?

**COAL CLINKERS:** Very light, porous, dark-colored and sharp-edged. Were these from coal-fired ships?

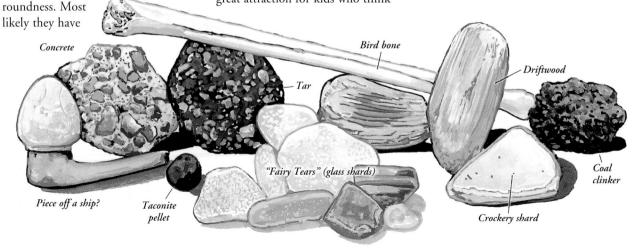

*Aluminum blob*

*Concrete* — *Bird bone* — *Tar* — *Driftwood* — *Piece off a ship?* — *Taconite pellet* — *"Fairy Tears" (glass shards)* — *Crockery shard* — *Coal clinker*

**BIRD BONES:** Often from gulls (Ring-billed and Herring) who fall prey to Great Horned Owls, Peregrine Falcons, storms, sickness and disease.

**CONCRETE:** Also know jokingly as "Portland conglomerate" (geologists think that's funny). Roads, rip-rap and concrete piers erode away giving the lake raw material to round and smooth into concrete "rocks."

**TAR:** Illegal dumping may account for some of the raw material for these man-made rocks. The lake then takes over, finishing the fragments into rounded nodules. You can break tar pieces in half with little effort.

**BRICK:** Tons of bricks were made from our wonderful clay that was left by Glacial Lake Duluth…

*Brick fragment*

and now many rounded fragments of brick are washing up on our beaches. They are usually yellowish, porous and very lightweight.

# BEACHES 101

The North Shore of Oahu and the North Shore of Minnesota do have something in common. They both have beaches and they both have surfers. (I once saw two guys surfing off the Knife River.) But the similarities end there. Let's talk about how our beaches came to be. Lake Superior's North Shore beaches are not the soft soothing sand beaches of the South Shore. They are rocky and rugged. How come?

### Source material

Sedimentary sandstone cliffs are the source for the South Shore's sand. The Apostle Islands are carved of sandstone. What the North Shore has to work with are banks of glacial till and solid cliffs of basalt and rhyolite. Waves break up cliffs and excavate the banks providing bite-size chunks of rock to a hungry and very restless Lake Superior.

Tumbling, cascading North Shore rivers — especially during spring runoff — contribute a fair amount of material to the lake as well. The lake counteracts by throwing the rocks back up forming ever changing rivermouth beaches (Figure 6). Come back tomorrow and you may not even recognize the place. Where you stood yesterday could be inundated. On more than one occasion I saw the Gooseberry River damned by rocks tossed up by the Lake overnight.

### Erosion: Breaking up is hard to do

How can little waves on a lake break up 200-foot rock walls? Remember, the "little waves" on Lake Superior can reach well over 30-feet and break

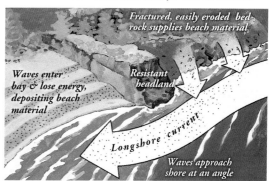

*Figure 5. A longshore current can supply a beach with material from fractured bedrock or old glacial moraine deposits.*

# BEACH FORMATION & GLOSSARY

Figure 6. River mouth supplying sediment to a beach.

700-foot ore boats in two. If you think about it, a good sized wave is packing thousands of tons of punch as it slams into a headwall. Measurements of Atlantic Ocean waves have recorded a force of 2000 pounds per square foot. Now that's impact. Pieces of the bedrock–especially at the weaker amygdaloidal/vesicular lava tops–are bound to break off. And let's not forget about the "silent breaker:" pressure. Wave water forced into cracks and cavities compresses the air inside. As the water recedes there is a rapid expansion of the air which can literally blast rock fragments from the wall. Once in the lake, the "new" sharp-edged stones, in an act of mutiny, turn on the bedrock they were spawned from and eat away at it via abrasion.

## Wave bye-bye

Lake Superior now has all the materials it needs to make a beach…It just needs to arrange them. To learn how the big lake does it, we must first know a bit about wave dynamics. Waves are round-topped swells in deep water. But when the bottom of a wave touches bottom it disrupts the circular flow and the wave crests (first stage of collapse). Waves usually attack the beach at the angle of the prevailing winds. As one end of a wave touches bottom, it slows down. The rest of the wave continues at the same speed until it hits bottom and collapses. The result is refraction: a gentle bending of the wave front. Since resistant cliffs jut out into the lake, they are encountered first when the wave energy is strongest. The graceful curve of the bay-beaches between points and headlands is the result of the loss of wave energy.

Material is also moved down the shore in a zig zag pattern. Waves come ashore at a slight angle but the backwash of sand, pebbles and rocks comes straight down the beach face. Literally tons of material is moved in this manner each year. Subtle but significant. Longshore currents also move rocks and pebbles (Figure 5).

## Portrait of a Beach

The first thing you may notice about a beach is the color of the rocks. Reddish beaches are made up of rhyolite, most likely from local lava flows. Multicolored beaches may be composed of a mix of local bedrock and glacial material brought south from Canada.

Take note of ridges of beach rock

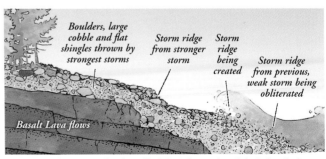

Figure 7. Lake Superior beach profile. The Lake giveth and the Lake taketh away.

running parallel to the shore. These are storm ridges and represent the high point of the wave energy from previous storms (Figure 7). Storm ridges are prevalent on the long sweep of Paradise Beach.

Large cobble high up on the beach may at first seem confusing. Wouldn't large rocks be closer to the water since they are harder to move? Yes and no. They stay put until a big storm throws them higher up the beach. There they sit until even bigger waves move them further back. Some beaches are composed of mainly flat rocks or shingles. Rounder rocks tend to roll back down the beach face leaving the harder to move flat shingles higher up the beach.

Whatever shore you are picking rocks on, remember that beaches are alive and ever changing. Return often and you will be rewarded.

*For more information on beach formation on Lake Superior see John Green's article in the magazine* **Natural Superior** *Volume 1: Number 1 and his book* **Geology on Display: Geology and Scenery of Minnesota's North Shore State Parks.**

# GLOSSARY

**Amygdule**. Mineral nodule in a vesicle (petrified gas bubble). They are created by seeping mineral-rich water that finds its way into an empty cavity and crystallizes. Quartz, agate, calcite and zeolite minerals are common amygdule types.

**Canadian Shield**. Name of the massive glacier-ground slab of granite bedrock that underlies much of eastern Canada, northern Minnesota, northern Wisconsin, Upper Peninsula of Michigan and northern New England.

**Cleavage**. Visible planes of fracture in a rock.

**Conchoidal fracturing**. Half-moon-shaped divots characteristic of quartzes such as agate, chert, flint, jasper and chalcedony.

**Crystal**. The form a mineral solution takes when it solidifies.

**Differential weathering**. When two rock types erode at different speeds leaving the harder rock jutting out beyond the softer rock. This is the process that creates waterfalls.

**Geode**. A crystal-lined hollow inside a rock. Quartz crystals are the most common geode lining.

**Geologist**. A person who studies rocks, minerals and earth processes.

**Glacier**. A humongous mass of ice that moves.

**Ice Age**. The time period from two million years ago to 10,000 years ago that spawned great grinding glaciers which advanced and retreated several times over the northern hemisphere.

**Igneous rocks**. Rocks formed by the earth's cooling magma (either underground or on the surface). Examples include basalt, rhyolite, diabase, granite and gabbro.

**Lava**. Hot liquid rock that flows out onto the earth's surface.

**Loupe**. A small but powerful magnifying glass worn around the neck which is used for close inspections of rocks.

**Magma**. Liquid rock found under the earth's surface.

**Matrix**. A fancy word for the main mass of a rock. Example: Most Thomsonite nodules formed in a matrix of basalt.

**Metamorphic rocks**. Igneous or sedimentary rocks that have been altered by heat or pressure. Examples are schist (was basalt or shale), marble (was

# Glossary & Index

limestone), slate (was shale), quartzite (was sandstone) and gneiss (was granite).

**Microcrystalline**. Extremely fine-grained quartzes. Examples are agate, chert, flint, jasper, chalcedony and carnelian.

**Midcontinent Rift Valley**. The now buried result of an attempt by the North American continent split 1.1 billion-years ago. Lava spilled out along the rift from Lake Superior to Kansas.

**Ophitic-textured**. A mottled, fuzzy appearance on the surface of basalt. Best seen when the rock is dry.

**Outcrop**. Exposed bedrock. Could be in the form of cliff, ridgetop, slab or ledge.

**Oxidization**. Change in a mineral when exposed to air.

**Phenocrysts**. Entire crystals of a mineral that hardened before the surrounding lava solidified. Porphyry is an example of a rock with phenocrysts.

**Plate tectonics**. The theory that the earth's crust is made up of several rigid segments or plates that are moving apart (rifting) or colliding (subduction).

**Pleistocene epoch**. The period from two million years ago to about 10,000 years ago. Also known as the Great Ice Age.

**Rift**. A gap in the earth's crust that oozes lava onto the surface. Usually due to an upwelling of magma from deep within the earth.

**Sedimentary rocks**. Rocks formed of particles eroded from other rocks. Examples are sandstone (usually quartz sand), shale (mud) and siltstone (fine-particled mud). Limestone is also a sedimentary rock but is formed of precipitated calcium carbonate.

**Storm ridges**. Terraces of beach rock formed by storm waves. One beach may have several visible storm ridges running parallel to the lake's edge.

 striations

**Striations**. Scratches or minute lines, generally parallel, inscribed on a rock or crystal surface.

**Stromatolites**. Fossilized algae found as wavy bands in Mary Ellen jasper.

**Till**. Material collected, carried and dropped by glaciers. Ranges in size from huge boulders to silt, clay and sand.

**Tombolo**. An island connected to the mainland by a beach. Grand Marais' Artist's Point is a classic tombolo.

**Vesicle**. Petrified gas bubble. A carbon dioxide or steam bubble trapped in a cooling lava flow (see amygdule). Found in basalt or rhyolite lavas.

# INDEX

**A**

agate, 1, 2, 3, 4, 5, 6, 7, 8, 11, 19, 22, 24, 25, 25-29, 31
— black, 27
— brecciated, 27
— crystal impression, 27
— eye, 27
— geode, 27
— gray, 27
— paintstone, 26, 27
— Paradise Beach, 7
— peeled, 28
— sagenite, 28
— shadow, 28
— skip-an-atom, 28
— stalactite, 29
— tube, 29
— waterlevel, 29
— waterwashed, 29
Alaska, 18
albite twinning, 34
amethyst, 24, 27
aluminum blob, 1, 3, 35
amygdaloidal, 1, 2, 3, 4, 5, 6, 9, 11, 14, 17, 22, 23
amygdules, 21, 22, 24, 30
— pipe, 11, 21
anorthosite, 34
Apostle Islands, 36
arrowhead, 32
Artist's Point, 6
asphalt, 4
Atlantic Ocean, 37

**B**

banded flint, 3, 4, 7, 32
banded rhyolite, 1, 17, 18
Baptism River, 5
basalt, 1, 2, 3, 5, 6, 9, 11, 12, 13-14, 15, 16, 17, 19, 20, 22, 25, 28, 30, 34
beach formation, 36-38
Beaver Bay, 4, 28
Beaver River, 4, 5, 18, 24, 32
biotite, 15, 16, 18, 19
bird bone, 35-36
black agate, 27
Boundary Waters Canoe Area, 18, 19
brecciated agate, 27
brick, 1, 2, 3, 36
Brooks Range, 18
Burlington Bay, 2, 3, 32

**C**

calcite, 3, 22, 23, 24, 27
calcium, 13, 16, 17
Canada, 11, 19, 26
Canadian Shield, 18, 19
Canal Park, 12
carbon dioxide, 9, 11, 20, 25
Carlton County, 26
Carlton Peak, 34
carnelian, 4, 7, 31-32
chalcedony, 2, 24, 28, 29, 31
— clear, 4, 32
— yellow, 7, 32
— smokey, 3, 32
chert, 8, 22, 31-32
— tan, 32
— yellow, 2, 3, 4, 32
chlorite, 22
coal, 35
Cold Spring, Minn., 19
columnar joints, 11
conchoidal fracturing, 32
concrete, 1, 4, 35-36
Continent East, 9
Continent West, 9
copper, 22, 34
Corner of the Lake, 1, 16, 18
crystal impression agate, 27
Cut Face Creek, 5

**D**

Denali, 18
Devil's Tower, 23
diabase, 3, 5, 15
dinosaurs, 10
Duck, Long-tailed, 5
Duluth, 1, 2, 9, 11, 35

**E**

earthquakes, 21, 28
El Capitan, 18
epidote, 22
eye agate, 27

**F**

fairy tears, 1, 2, 3, 35
feldspar, 14, 18, 23, 27, 34
flint, 4, 24, 31-32

# INDEX

— banded, 3, 4, 7, 31-32
Flood Bay, 3, 15, 20, 24, 28, 32

**G**

gabbro, 1, 2, 3, 16
geode, 21, 24
geode agate, 27
Georgia, 34
Giant Camel, 12
Giant Short-faced Bear, 12
Glacial Lake Duluth, 12, 36
glaciers, 10, 11, 12, 13, 17, 19, 25, 26, 28
glass, 1, 35
gold, 24
Good Harbor Bay, 5, 6, 30, 33, 34
Gooseberry Falls State Park, 3, 4, 15, 18, 20, 32
Gooseberry River, 36
Grand Marais, 5, 6, 7, 18, 27, 30
granite, 1, 2, 3, 4, 7, 8, 18-20, 24
granitic magma, 17
gray agate, 27
Great Horned Owl, 36
Green, John, 10
Gull, Herring, 36
Gull, Ring-billed, 36

**H**

Half Dome, 18
Hawaii, 9, 13
hematite, 31-32
Herring Gull, 36
hornblende, 18, 19
Hovland, 7

Hudson Bay, 11

**I**

ice, 10, 11
Ice Age, 10, 11, 26
Iceland, 9, 13
igneous, 6
Illinoisan Glaciation, 10
Iowa, 9, 11, 26
iron, 13, 16, 17, 25-26, 27, 31, 32, 35
Iron Range, 28, 35
Isle Royale, 34

**J**

jasper, 2, 3, 4, 8, 24, 31-32
— Mary Ellen, 31

**K**

Kansan Glaciation, 10
Kansas, 9
Keewenaw Peninsula, 34
Kitchi Gammi Park, 1, 16, 18, 24, 32
Knife River, Minn, 34

**L**

Lake Superior, 2, 3, 4, 9-12, 13, 16, 25, 26, 34, 36
Lake Superior basin, 10, 11, 13, 25
lava, 6, 9, 10, 11, 13, 15-17, 19, 20, 21, 23, 25, 26, 33, 37
lintonite, 3, 34
Long-tailed Duck, 5
longshore currents, 37, 38

**M**

magnesium, 13, 16, 17
magnetite, 33
Mammoth, 12
Mary Ellen jasper, 31
Mastodon, 12
Matsch, Charles, 9
metamorphic, 11
Michigan, 10, 19, 34
microcrystalline quartz, 2, 4, 7, 25, 27, 31
Mid-Atlantic Rift, 9
Midcontinent Rift Valley, 10, 11, 19
Minnesota, 9, 10, 11, 19, 26, 34, 35, 36
*Minnesota's Geology*, 9
Montana, 34
Moose Lake, 25
Mt. McKinley, 18

**N**

Nebraskan Glaciation, 10, 11
New England, 19
New York, 34

**O**

Ojakangas, Dick, 9
Ojibwa Indians, 32
Oldsquaw, 5
olivine, 22
ophitic-textured basalt, 1, 2, 4, 6, 14
orthoclase, 18, 19, 23
Owl, Great Horned, 36

## P

paintstone agate, 26, 27
Paleo-Indians, 32
Palisade Head, 17, 18, 21, 23
Palisade Head Flow, 10, 21
Palisade rhyolite, 18, 23
Paradise Beach, 7, 20, 24, 32, 38
Paradise Beach agate, 7, 27
peeled agate, 28
Peregrine Falcon, 36
phenocrysts, 18, 23
pipe amygdules, 11, 21
pipe vesicles, 20
plagioclase, 14, 16, 18, 19, 23, 33, 34
Pleistocene, 25
porphyry, 4, 6, 17, 18, 22, 23
potassium, 13, 17
Precambrian Era, 13
prehnite, 22, 34
pyroxene, 33

## Q

quartz, 1, 3, 4, 5, 6, 7, 8, 11, 18, 19, 21, 22, 23, 24, 25, 27, 31-32
— microcrystalline, 2, 4, 7, 25, 27, 32
— rose, 24

## R

Rapp, 30
rhyolite, 1, 2, 3, 5, 6, 9, 11-13, 16, 17, 19, 22, 25, 32, 33
— banded, 1, 17, 18
— Silver-Beaver, 4, 5, 17, 18

rifts, 9, 13
Ring-billed Gull, 36
Rocky Mountains, 18
rose quartz, 24

## S

Saber-toothed Cat, 12
sagenite agate, 28
sandstone, 1, 4, 5, 6, 10, 11, 24, 33, 36
Sawtooth Mountains, 30
sedimentary, 6, 32, 36
shadow agate, 28
shale, 10, 11
sharks, 10
Shovel Point, 17, 23
Sierra Nevada Mountains, 18
silica, 13, 17, 27, 28, 29, 31
siltstone, 3, 32
Silver Bay, 3, 4, 35
skip-an-atom agate, 28
Skyline Drive, 12
Split Rock Lighthouse, 34
Split Rock River, 4, 14, 20, 32
stalactite agate, 29
Stoney Point, 1, 14, 20
storm ridges, 1, 7, 37, 38
stromatolites, 31
sulphur dioxide, 9, 20

## T

taconite pellets, 35
tar, 35-36
Terrace Point Flow, 30

Tettegouche State Park, 5, 17, 18, 34
Thomsonite, 5, 22, 30, 34
Thomsonite Beach, 5, 30
Thomsonite Beach Inn & Suites, 5, 30
tombolo, 6
tube agate, 29
Two Harbors, 2, 3, 28, 35

## U

Upper Peninsula of Michigan, 19, 34

## V

vesicles, 20, 22, 23, 25, 31
vesicular, 2, 3, 9, 11, 14, 17, 20, 23, 34
— pipe, 20
volcano, 23, 33

## W

Wallace, 30
waterlevel agate, 29
waterwashed agate, 29
Wisconsin, 11, 19
Wisconsin Glaciation, 10, 11
Wyoming, 23

## Z

zeolite, 20, 21, 30, 34

# Titles of Interest

Daniel, Glenda, and Sullivan, Jerry, *A Sierra Club Naturalist's Guide: The North Woods*. San Francisco: Sierra Club Books, 1981.

Gauthier, Kevin, *Lake Superior Rock Picker's Guide*. Michigan: Petoskey Co-Pub, 2007.

Green, John C., *Geology on Display: Geology and Scenery of Minnesota's North Shore State Parks*. St. Paul: Minnesota Department of Natural Resources, 1996.

LaBerge, Gene L., *Geology of the Lake Superior Region*. Tucson: Geoscience Press, 1994.

Lynch, Bob and Dan Lynch, *Lake Superior Rocks & Minerals*. Cambridge, Minnesota: Adventure Publications, 2008.

Lynch, Bob and Dan Lynch, *Agates of Lake Superior: Stunning Varieties and How They Are Formed*. Cambridge, Minnesota: Adventure Publications, 2011.

Lynch, Dan, *Lake Superior Agates Field Guide*. Cambridge, Minnesota: Adventure Publications, 2012.

Magnuson, Jim, *The Storied Agate: 100 Unique Lake Superior Agates*. Cambridge, Minnesota: Adventure Publications, 2011.

Magnuson, Jim, *Agate Hunting Made Easy: How to Really Find Lake Superior Agates*. Cambridge, Minnesota: Adventure Publications, 2012.

Ojakangas, Richard W. and Matsch, Charles L., *Minnesota's Geology*. Minneapolis: University of Minnesota Press, 1982.

Polk, Patty, *Collecting Agates and Jaspers of North America*. Iola, Wisconsin: Krause Publications, 2013.

Pough, Frederick H., *Peterson Field Guide: Rocks and Minerals*. Boston, New York: Houghton-Mifflin Company, 1988.

Roberts, David C., *Peterson Field Guide: Geology Eastern North America*. Boston, New York: Houghton-Mifflin Company, 1996.

Sansome, Constance Jefferson, *Minnesota Underfoot*. Stillwater: Voyageur Press, 1983.

Sorrell, Charles A., *Minerals of the World: A Golden Guide*. New York: Golden Press, 1973.

Wolter, Scott F., *The Lake Superior Agate*. Edina, Minnesota: Burgess Publishing, 1994.

Wolter, Scott F., *Amazing Agates: Lake Superior's Banded Gemstone*. Duluth, Minnesota: Kollath-Stensaas Publishing, 2010.